Germ Stories

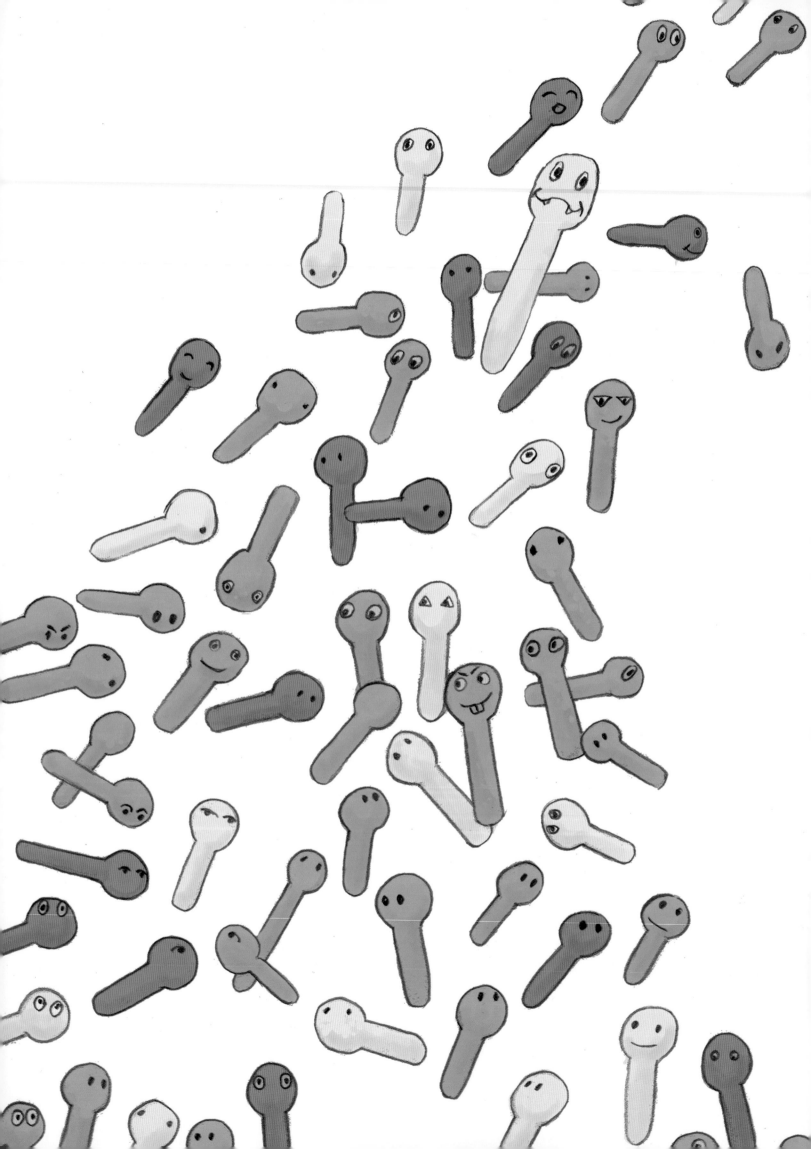

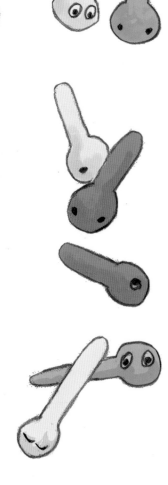

Germ Stories

Arthur Kornberg

ILLUSTRATIONS BY

ADAM ALANIZ

PHOTOGRAPHY BY

ROBERTO KOLTER

UNIVERSITY SCIENCE BOOKS   SAUSALITO, CALIFORNIA

University Science Books
www.uscibooks.com

Order information:
  Phone: 703-661-1572
  Fax: 703-661-1501

Library of Congress Cataloging-in-Publication Data
Kornberg, Arthur, 1918–
   Germ stories / Arthur Kornberg ;
illustrations by Adam Alaniz ;
photography by Roberto Kolter.
   p. cm.
   ISBN 978-1-891389-51-1 (alk. paper)
   1. Microbiology—Juvenile literature. 2. Bacteria—
Juvenile literature. I. Alaniz, Adam, ill. II. Kolter, Roberto,
1953– III. Title.
   QR57.K67 2007
   616.9'041—dc22   2007009960

Printed in Hong Kong

To all, young and old,

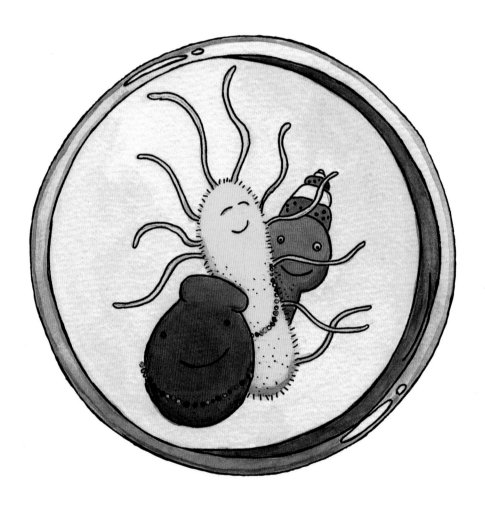

who adore "the little beasties"

Contents

A colony of the spore-forming
Bacillus subtilis.

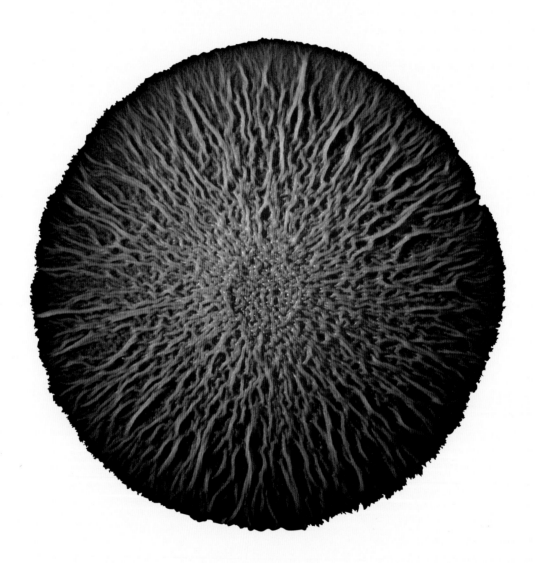

 The tip of a flathead screwdriver is almost as
long as 10,583 colonies of *Bacillus subtilis.*

Preface

I told my three sons stories about germs more than fifty years ago as fanciful bedtime tales. Whatever substance the tales had came from three sources: my training in clinical medicine, my undertaking to remake a classical medical school department of bacteriology with exclusive emphasis on disease germs, and, finally, my attending the extraordinary course on microbiology given by C. B. van Niel at the Hopkins Marine Station in Pacific Grove, California. Only the good "little beasties" were studied, and even the mention of a pathogenic microbe was absolutely forbidden.

Years later, on the several occasions when I took one and then another of my eight grandchildren on an extended lecture trip, their fathers urged that I tell them the *Germ Stories*. No longer able to concoct such tales, I instead tried to make instructive and entertaining rhymes and included in each poem a grandchild's name or that of a cousin.

This collection of stories, written more than ten years ago, was circulated among friends and family until Bruce Armbruster of University Science Books undertook to publish it. The only additions are three irresistible stories: one is about *Helicobacter pylori*, the ancient and still significant inhabitant of the stomach; another describes the voracious, predatory behavior of *Myxobacterium xanthus*; and the third, "The Intestinal Menagerie," gives an account of the unique bacterial population each human has in many trillions that exceed tenfold the cells in the body. A significant revision is the final couplet of the AIDS story. Current medical treatment, due to advances in science, has changed AIDS from a fatal disease to one that is manageable. It is a happy note on which to end that poem.

To Bruce and to the staff of Wilsted and Taylor, particularly Christine Taylor and Melody Lacina for their skillful copyediting, and to Adam Alaniz for the illustrations and Professor Roberto Kolter for the photography, I am deeply grateful.

Germ Stories

The Germ Parade

Hurry, hurry to the parade
Of the strangest creatures ever made.

No legs, no fins, no mouths, no eyes,
Little beasties of the tiniest size.

Far too small for the eye to see—
"Just how small *is* this menagerie?"

Imagine, Zac, if you can,
A tiny dot, a grain of sand.

Break each grain into tinier ones still—
Into a thousand, if you will.

Into each minigrain (big enough),
Thousands of germs you can stuff,

With lots of room for every germ
To swim and tumble, turn and squirm.

Want to see them, every kind?

"Where to look, and how to find?"

Everywhere! In soil and air—
They're on your skin, your nails, your hair.

From between your teeth, scrape out some goo
Or take some dirt from off your shoe.

Spread it on a clean glass slide,
Under a microscope magnified.

Now peek through the lens, and in the light
A new world appears. Fantastic sight!

Rods, short and long, dart in and out,
Among dead stuff they weave about.

Germs can be wispy or thick or round,
All alone or in groups they are found.

"Wow! A hairy monster is swimming!
And there's a snaky thing that's wriggling!

"Can these germs live inside me?
In dogs and cats? In a fish? In a tree?"

Yes, Zac. It's unquestionably true,
Within your bowels there is quite a zoo.

Some germs are helpful, really good guests,
While others can hurt you. They are the pests.

**Now I'll describe germs good and bad,
And boys and girls, made sad or glad.**

Staphylococcus Aureus

(Staf·lo·kok´·us Or´·ee·us) **FOOD POISONING**

Staph aureus is on your hands and on your hair.
It's in your nose—it's everywhere!

If you prick your skin, it enters and thrives:
Millions of germs, very much alive.

But our body cells and antibodies
Can vanquish invaders with the greatest of ease.

When germs are surrounded, they soon are gone.
The battle is fought and quickly won.

From a baker's hand on one rare occasion
Some *Staph* broke out for a new invasion,

This time into a warm custard pie
Baked that day. Oh, dear. Oh, my!

In the pie germs grew and frolicked

And spewed out poisons that can cause colic.

Along came Jessica home from school.

When she saw the pie, she began to drool.

For dinnertime she couldn't wait

And from the pie she ate and ate

Far more than a little girl should.

She couldn't stop, it was so good.

Late that night as she lay in bed

Her tummy ached. So did her head.

To the bathroom she had to run
Again and again, it was no fun.

"Mom—I'm hot, I'm cold, the room's awhirl."
"Oh, Jessie darling, my poor little girl.

"Doctor Jacobs, please do come quick.
Our Jessie dear is very sick."

"I know the problem," the doctor said.
"**Food poisoning** is what I dread.

"To the hospital! She *must* go there
For fluids, shots, and nursing care."

After some days Jess began to heal,
Hungry again for a good big meal.

But one thing you can know for sure,
Custard pie held no allure.

And now you know:
when handling food,
Wash your hands,
germs to exclude.

Staphylococcus aureus: A spherical bacterium commonly found on skin and mucous membranes. When colonies of *Staph aureas* are grown in a medium containing blood, they break apart the red blood cells, producing clear halos around the colonies, as shown in this photograph.

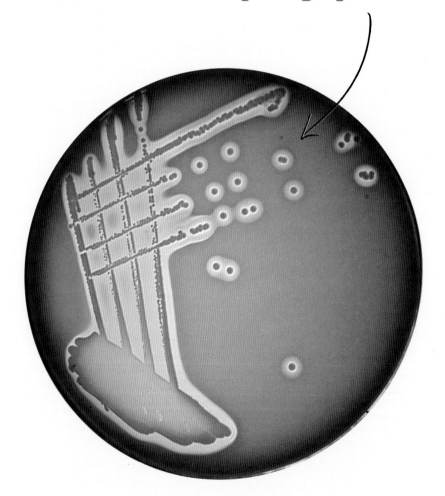

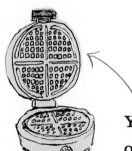

You could fit about 22,473,516,200 colonies of *Staph aureas* in a single hole of your waffle.

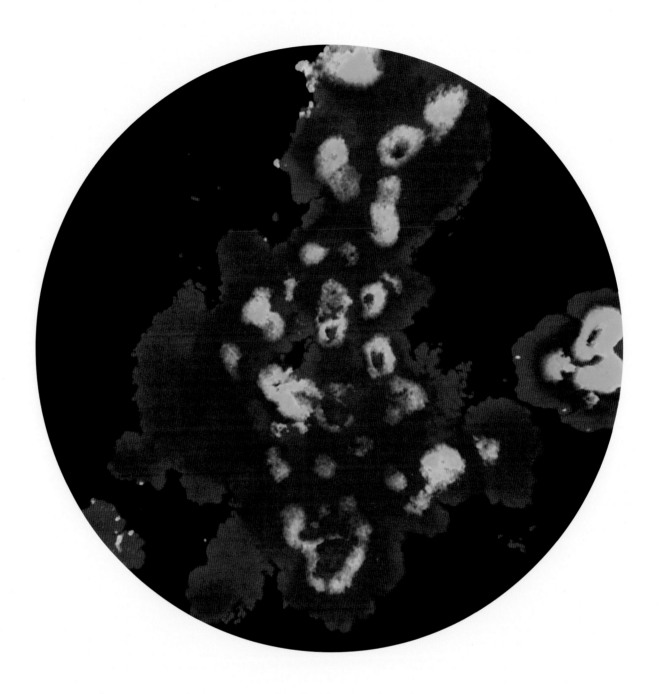

Staphylococcus aureas growing on a surface in a biofilm. Its grapelike clusters, which appear golden as a colonial mass, can produce toxins that invade the body and cause serious infections.

Salmonella Typhi

(Sal·ma·nel´·a Tie´·fee) **TYPHOID FEVER**

"Sophie, you haven't touched a bite."

"Sorry, Mama: no appetite."

"Off to bed, you're tired and blue."

"My tummy aches; my head does, too."

Next morning Sophie felt like jelly.

"Do not press my tender belly!"

Her temperature: one hundred four!

And her body showed red spots galore.

What Sophie had was quite obscure.

The doctor frowned. Since he wasn't sure,

He sent her off with ambulance din

For lab tests, nursing, and medicine.

In the tests it was plain to see

Sophie's germs: **Salmonella typhi.**

Ill for weeks with **typhoid fever**,

She was given antibiotics so it would leave her.

"From where," she asked, "came these bad *typhi*?

A place remote that I could not see?"

"Sophie, dear, try hard to think.

Did you ever take strange food or drink?"

Sophie thought and scratched her head.

Was there a clue? Where had she tread?

"I remember once at a certain spot

I was tired, thirsty, and very hot.

"In the creek on the way from school,

The water looked clear and fresh and cool.

"I took a sip, and it tasted so good,

I kept on drinking as much as I could."

They found the creek, and up the hill

There lived a lady who filled the bill.

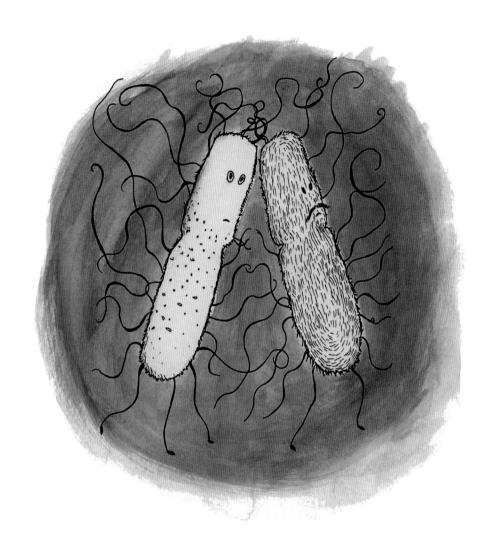

Recovered from typhoid, she was not fully cured,
so **Salmonella typhi** in her endured.

From her small home a sewer leaked,
And in its flow, germs reached the creek.

Beware food or drink away from home.
Bad germs may lurk wherever you roam.

Typhoid carriers: be fully healed!
And please be sure your pipes are sealed.

This photograph shows a human cell with
several **Salmonella typhi** cells surviving inside.

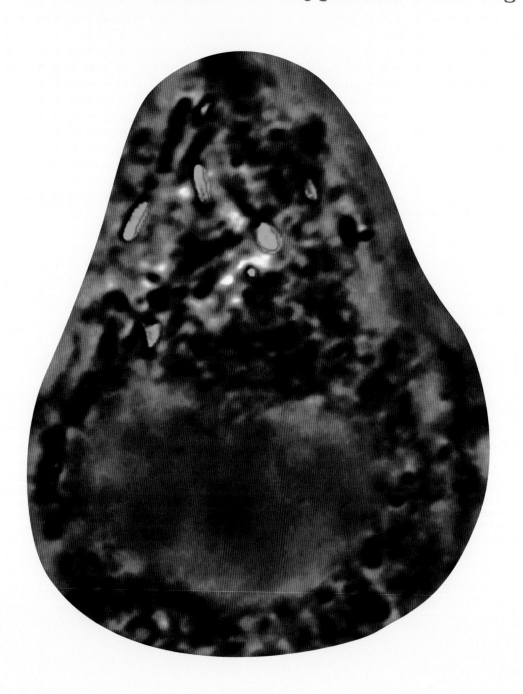

Salmonella typhi: A rodlike bacterium related to the many bacteria that normally reside in the intestines. The toxins it produces invade the bloodstream to cause typhoid fever. The photograph below shows *Salmonella typhi* grown as streaks on a nutritive agar plate.

 How big is a penny? It's about the size of 283,515,560 *Salmonella typhi*.

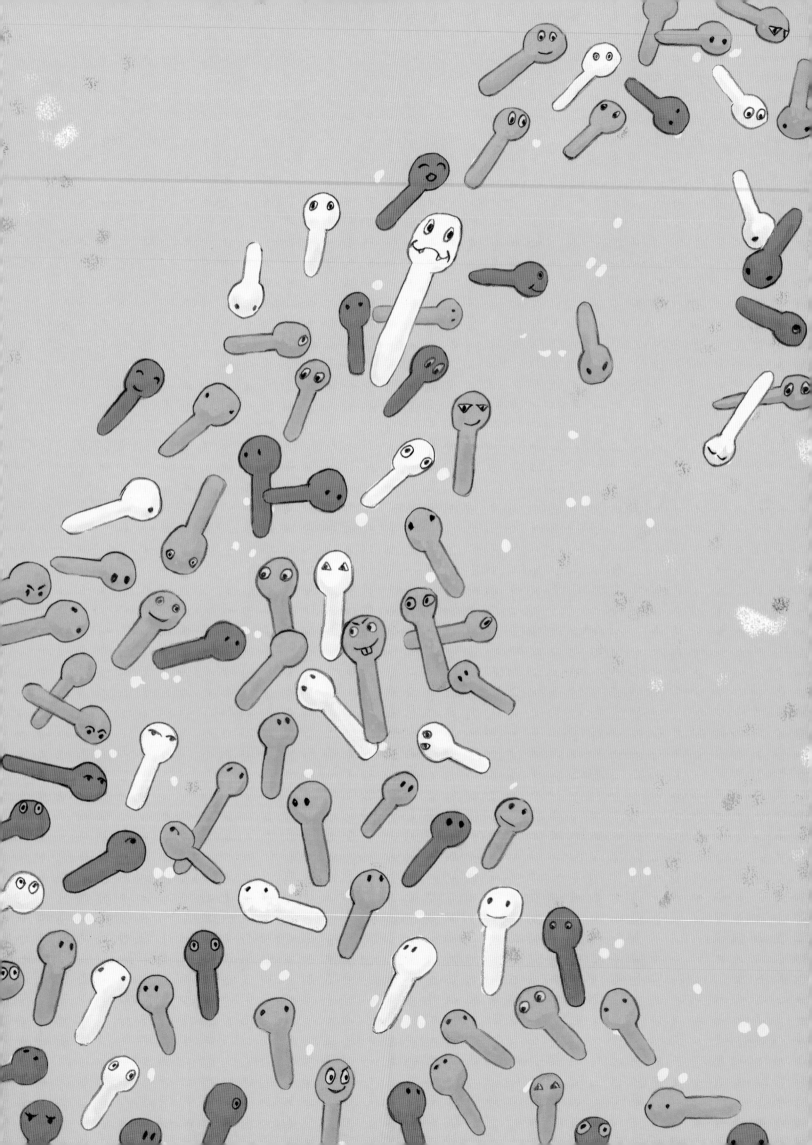

Clostridium Tetani

(Klos·tri·dee´·um Tet´·an·eye) TETANUS

This germ with many coats around

For years will slumber in the ground.

As a spore it can be boiled or frozen,

And it resists toxins while it's a-dozin'.

Should chance allow the spore access

To a cozy place that's free from stress,

The coats are shed. When moist and fed,

Clostridium emerges, grows, and spreads.

When barefoot Gili stepped on a nail,

Then began our sad spore tale.

A little sting, a break in the skin—

That's how dreadful tetanus begins.

Inside his heel in a silent state,

The spore began to germinate.

It stirred and cast its coats aside.

Into two, into more it multiplied.

Clostridia can grow so fast

That in a day, millions of germs are amassed.

The poisonous toxins they liberate

Throughout the body circulate.

Poor Gili cried all through the night.

When morning came his jaws were tight.

Muscles so stiff he couldn't walk:

He couldn't swallow, couldn't talk.

"Nine one one! Emergency! Please!

Gili is blue. He has a disease."

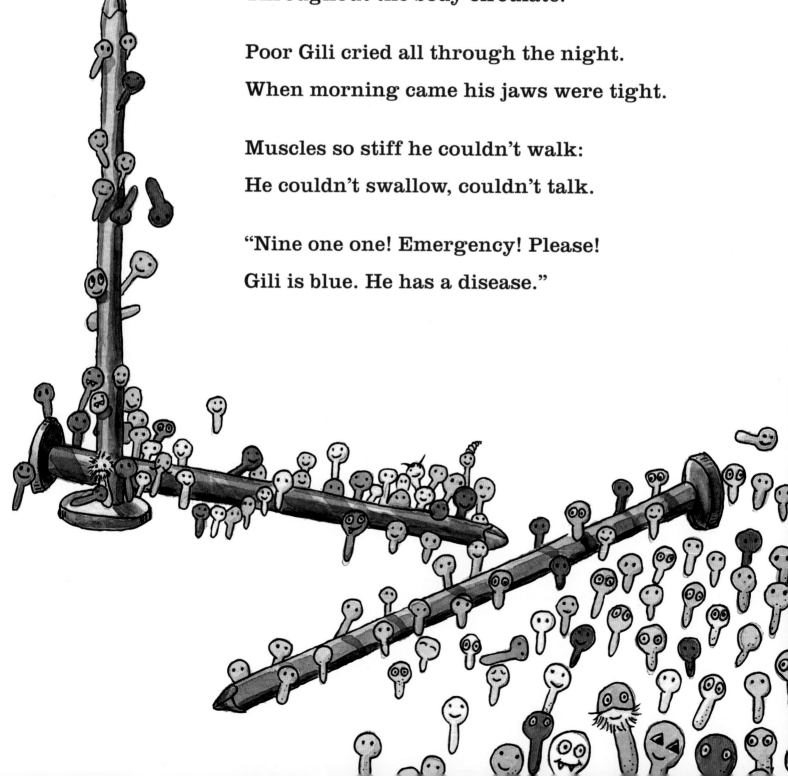

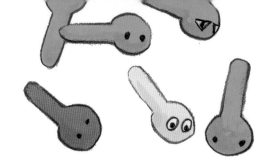

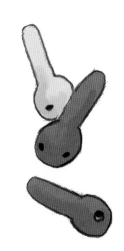

The doctor knew. "It's tetanus.
This germ's toxins are dangerous.

"A lung machine will help him breathe,
And then two drugs can intervene.

"Antitoxins will make the spasms relax,
Antibiotics will give those germs the axe."

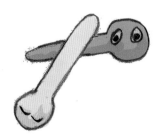

With toxins by antitoxins bound,
Gili's muscles grew soft and round.

Antibiotics halted **Clostridia's** fast growth
By messing up the germs' protective coats.

**Gili is healthy, but now we know
That body defenses can get too low.**

**Even though you've had a tetanus shot,
A booster when you're injured helps a lot.**

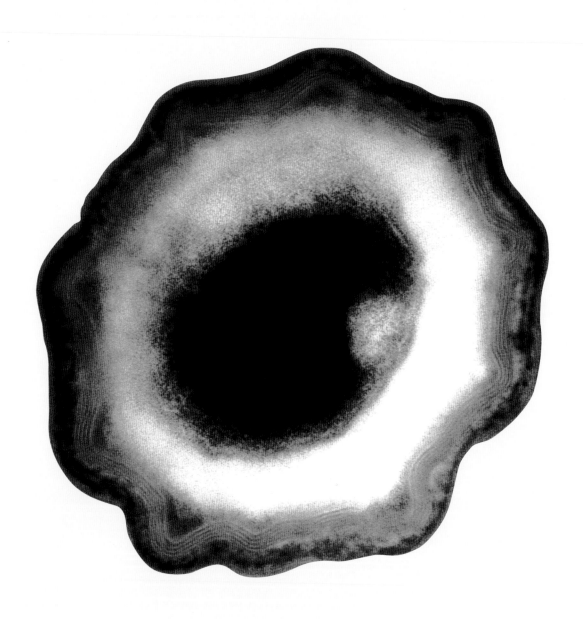

Clostridium tetani: A rodlike bacterium that can become a spore to attain a highly resistant dormant state. When introduced through a wound, the spore can germinate to become a bacterium again, releasing toxins that cause severe muscle contraction in the arms, legs, diaphragm, and jaw. This electron micrograph of a dormant spore shows its many protective layers.

44,000—or more—spores of *Clostridium tetani* could fit on the tip of a pushpin. Watch out.

Saccharomyces Cerevisiae

(Sa·kar´·o·my·sees Ser·a·vee´·see·yay) **YEAST**

Guy, remember this name today:
Sa-kár-o-my-sees ser-a-vée-see-yay.

A very good germ for many reasons,
It's helpful to us in all the seasons.

This tiny **yeast** from grapes makes wine
And brews the beer that fills the stein.

The same gas bubbles that help dough rise
Give sparkling wine its sweet surprise.

A thousand years a mystery—
It's now understood through chemistry.

The sugar molecule when rearranged
Into *two* molecules is changed.

One, alcohol, is a great elixir

In medicine and cocktail mixers.

The other, carbon dioxide (CO_2),

Forms the heady foam of a tasty brew.

"How are all these things brought about?"

Yeast enzymes do it, without a doubt.

Enzymes, like chemists in the cell,

In these operations do excel.

So cakes of yeast are sold for baking,

And tons of yeast used for beer making.

"Why," Guy asked, "does yeast take the trouble

To make alcohol and those gas bubbles?"

By splitting sugar it makes energy
That it can store as ATP.

"What is ATP?" asked Guy.
Cells use it to grow and multiply.

Little buds from yeast cells grow
And if fed ATP will keep doing so.

"What's a cell? How does it eat?
Does it have a mouth and teeth?"

No, no, **Saccharomyces** is a tiny ball.
It has no outside parts at all.

It takes in food through its thin skin.
Enzymes digest the food within.

The cells of plants, animals, and us
   Are very many and various.

But deep inside, differences are least
   Between cells of humans, yeast, and beast.

This electron micrograph of a starved yeast cell shows the compartments (vesicles) that usually store food.

Saccharomyces cerevisiae: A yeast fungus found in fruit or soil or on human skin. It is widely used in baking bread and brewing. This photograph of a yeast colony growing on a hard agar surface shows very elaborate stalks.

It would take at least 8,000,000,000,000 yeast cells to fill your nostril. Yuck.

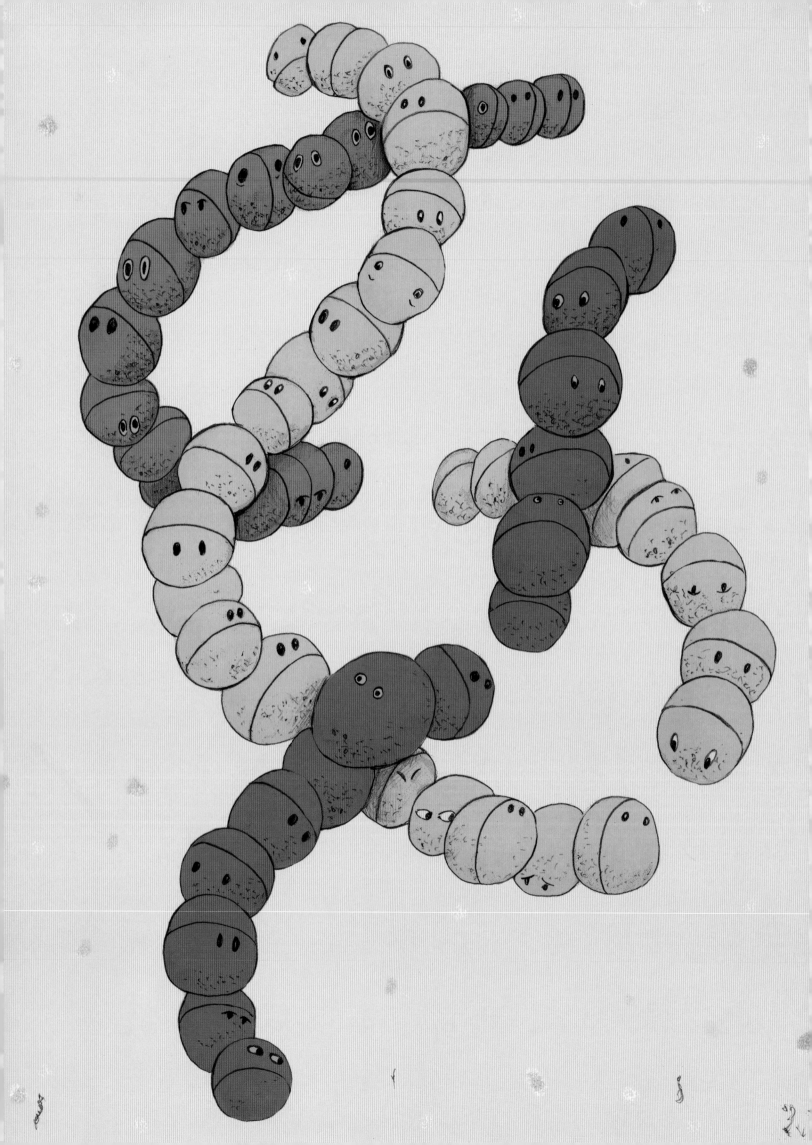

Streptococcus Pneumoniae

(Strep·to·kok´·us Noo·mo´·nee·ay) **PNEUMONIA**

Ross had fever, aches, and chills.

He had the flu but wasn't that ill

Until he began to cough much more,

And his fever rose to one hundred and four.

"Penicillin and rest in bed

Should cure him soon," the doctor said.

Alas, things took another turn.

His lungs became the main concern.

Gurgles were heard with stethoscope.

But still the doctor held out hope

That penicillin might hold at bay

The **Streptococcus** *noo-mo-nee-ay.*

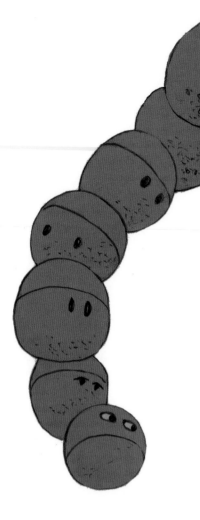

Penicillin stops germs from making cell walls.

With leaky walls germs can't grow at all.

But a gene in one of Ross's germs mutated

So its fast growth was not abated.

The mutant cell could penicillin split

And thus become quickly rid of it.

The resistant mutant then multiplied

Until a whole lung lobe was occupied.

"We must give Ross another drug

To stop the growth of this resistant bug.

"Which antibiotic would be the best

To kill the mutant in a test?"

Cephalosporin withstood the blow

And did not allow the mutant's wall to grow.

"We'll give Ross cephalosporin by vein.

Let's hope no mutant germs remain."

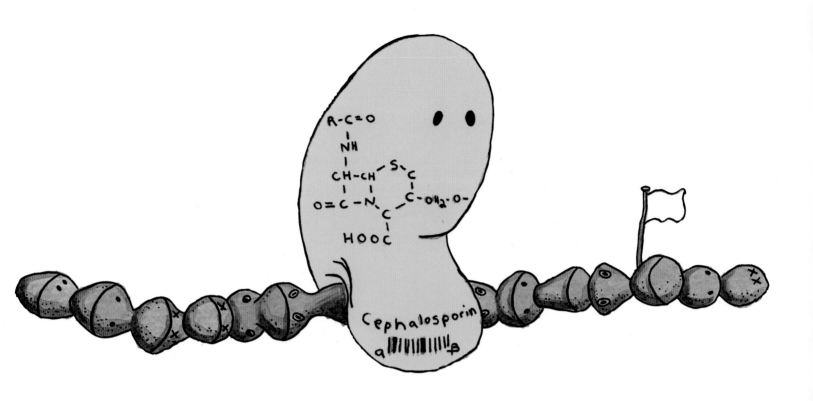

Ross's fever fell within a day,
His cough and aches soon went away.

His lungs expanded without pain.
Ross was playful and happy again.

While a cold or flu is not much ado,
Take heed of what it may lead to.

Bad cough, chest pain, and fever high—
Pneumonia germs may be close by.

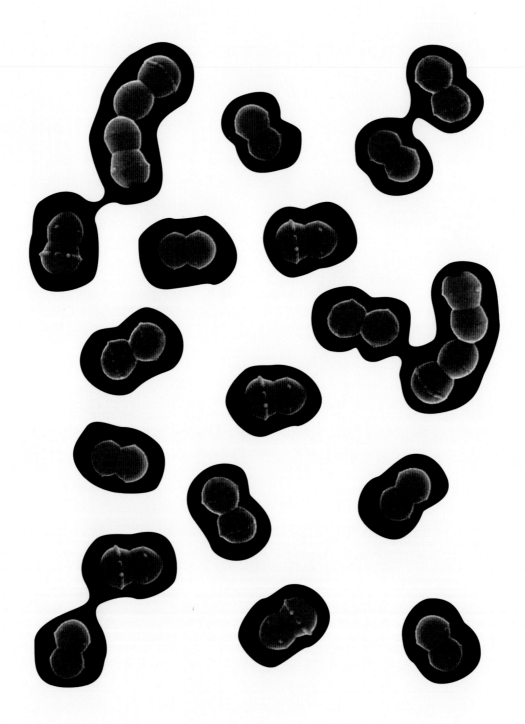

Your fingernail is almost 75,480,700 times larger than a single *Streptococcus pneumoniae* bacterium.

Streptococcus pneumoniae: A small, round bacterium found in pairs or short chains, as in the facing photograph. Strep causes serious human diseases such as lobar pneumonia. The electron micrograph below shows the bacteria growing in a biofilm on glass.

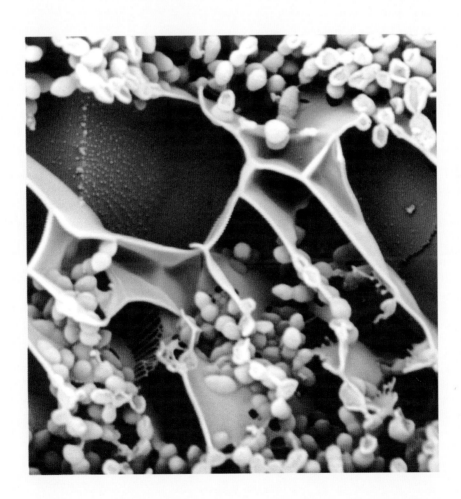

Myxobacterium Xanthus

(Miks·o·bac·teer´·ee·um Zan´·thus) **GERM WARS**

Named *Myxo* for mucus and *xanthus* for yellow,
This voracious predator is anything but mellow.

Each Myxo spews slime on which it can glide.
It devours bacteria. It's a microbicide.

Hunting in "wolf packs," the Myxo stalks prey.
With many hungry mouths, it digests victims away.

"Where are the Myxo?" Zoe asked her granddad.
"And where are the bacteria that make them so mad?"

Thriving in soil until all bacteria are gone,
Myxo rods then collect to await a new dawn.

A hundred thousand rods are in each **Myxo** mass.
Each rod becomes a spore that no stress can harass.

Sleeping for years, the spores thus await

Nutrient signals nearby bacteria will someday generate.

Then the spores will awaken, shed their coats, and grow,

Forming hungry "wolf packs" again, as predatory foes.

Zoe was curious and wanted to know,

"What's inside **Myxo** that makes them just so?"

It's the fascinating molecule, the great poly P,

Which chemistry can explain in a future germ story.

A swarming colony
of *Myxobacterium xanthus*.

Myxobacterium xanthus: One of many species of complex bacteria that produce slime on which they glide to find other bacteria and feed on them. This electron micrograph shows *Myxobacteria* growing on decaying forest wood. Its fruiting body has dormant spores at the tips.

Speaking of fruit, about 2,093,300 fruiting bodies of *Myxobacteria* could cover the skin of a blueberry.

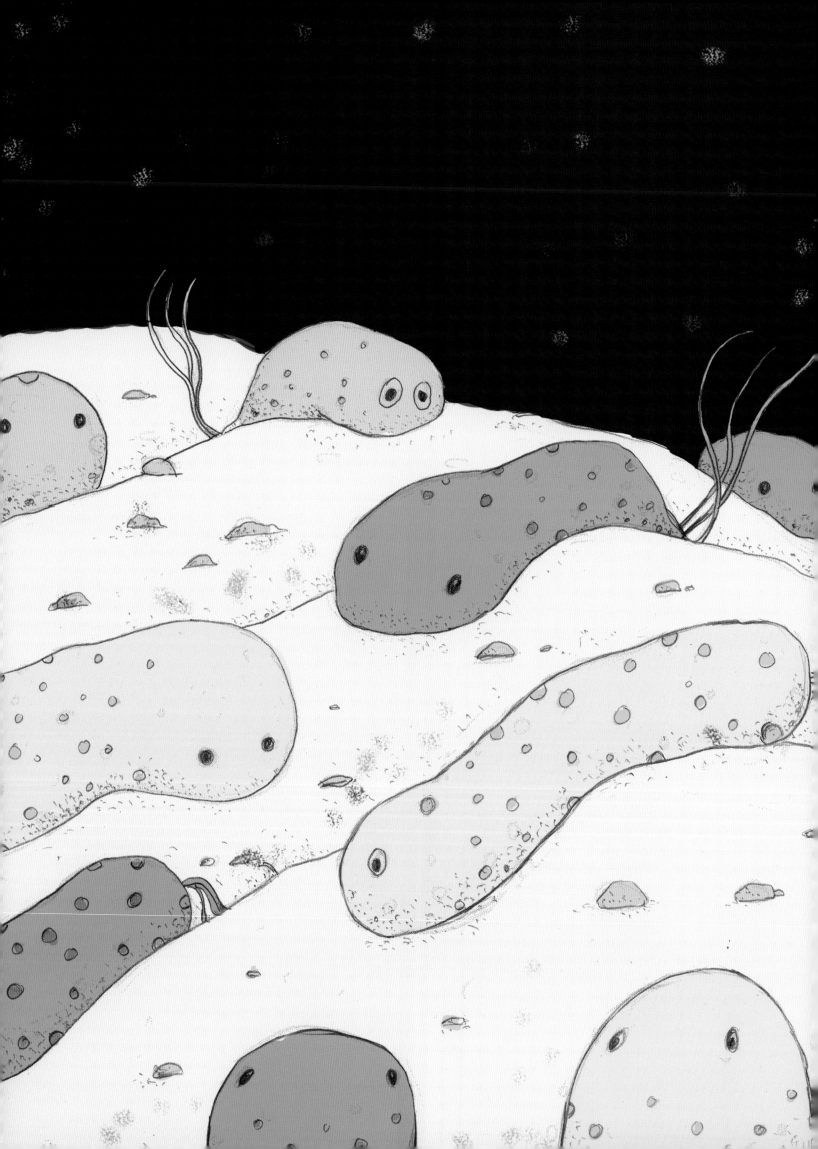

Helicobacter Pylori

(Hee´·li·ko·bak·ter Pie·law´·ree) **STOMACH ULCER**

This corkscrew bug thrives among us
By burrowing into the pylorus.*

Corrosive stomach acid the bug withstands
By making an enzyme that renders it bland.

This bug's been found in each hemisphere,
In all kinds of people for ten thousand years.

In some, *H. pie-láw-ree* makes peptic ulcers
That left untreated form stomach cancers.

To doctors those people seemed fraught with anxiety,
Which was not relieved by antacids or psychiatry.

"It seems strange to me that a germ attack
Wasn't thought of then," said little Zac.

*The pit of the stomach.

Barry Marshall was the one who thought **H. pylori**,

This most common bug, could be so contrary.

In 1984 that researcher courageous

Swallowed lots of *H. pylori*.

Said Zac, "That could be dangerous!"

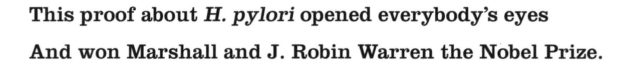

When the bugs made him sick,

An antibiotic was the cure.

Marshall proved *H. pylori* guilty.

Of that we are sure.

This proof about *H. pylori* opened everybody's eyes

And won Marshall and J. Robin Warren the Nobel Prize.

With antibiotics now in such wide use,

That contrary **H. pylori** has begun to snooze.

But without *pylori* enzymes, stomach acid has reappeared,

And these days, acid backup into the esophagus is feared.

There bad acid can cause cancer anew.

"How can we prevent it?" asked Zac.

We're not sure yet what to do.

Helicobacter pylori: A helix-shaped, common bacterium that screws into the lining at the pit of the stomach (the pylorus). It releases an enzyme that neutralizes corrosive stomach acid. In some it creates an open sore, a peptic ulcer, that may progress to cancer. This electron micrograph shows the flagella that allow *H. pylori* to swim.

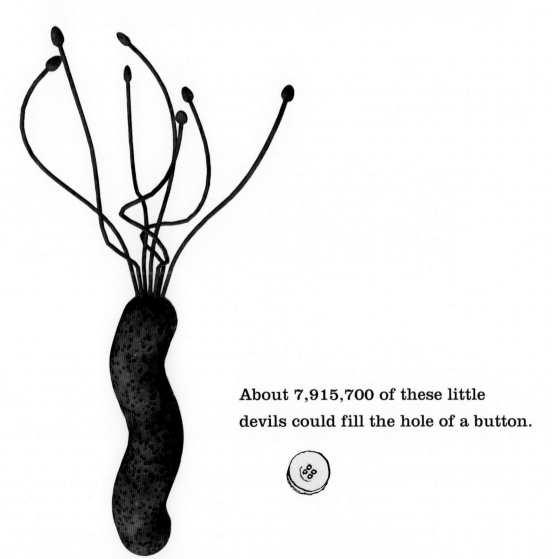

About 7,915,700 of these little devils could fill the hole of a button.

Penicillium Notatum

(Pen·i·sill´·ee·um No·tah´·tum) **PENICILLIN**

"What's penicillin and where's it from?"
Maya asked her dad and mum.

"Why is it good medicine for me?
How did it kill the germs in Sophie?"

This is the story that they told
About the truly astonishing
Penicillium mold.

Penicillin is made
By the **Penicillium** mold
To kill bacterial germs
That have gotten too bold.

The walls of bacteria are unique, not weak,
But penicillin attacks them and makes them leak.

Because leaky cells
Cannot survive,
No bacteria
Can remain alive.

Since human cells don't have such walls,
To them, penicillin is no bother at all.

Said Maya, "This is all so confusing.
Is mold the medicine doctors are using?"

No, Maya, it's not like that.

The penicillin drug is made in a vat.

"I want to see bacteria and watch them grow.

How penicillin kills them I'd like to know."

Maya, bacteria are too small to see.

But when millions of them form a colony,

They make a spot of sand-grain size

That is quite easy to recognize.

When many colonies are together drawn

They form a mat, a bacterial lawn.

Should a **Penicillium** mold among them be,

Its penicillin would clear a zone you could see.

If on the mat a spot is clear,

The *Penicillium* mold is saying, "I was here."

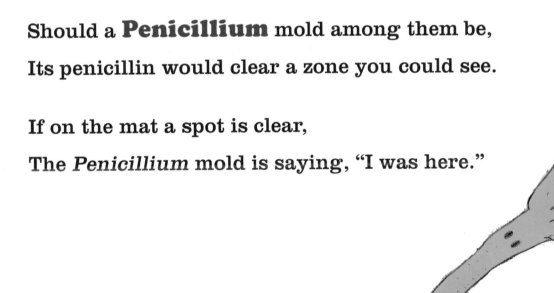

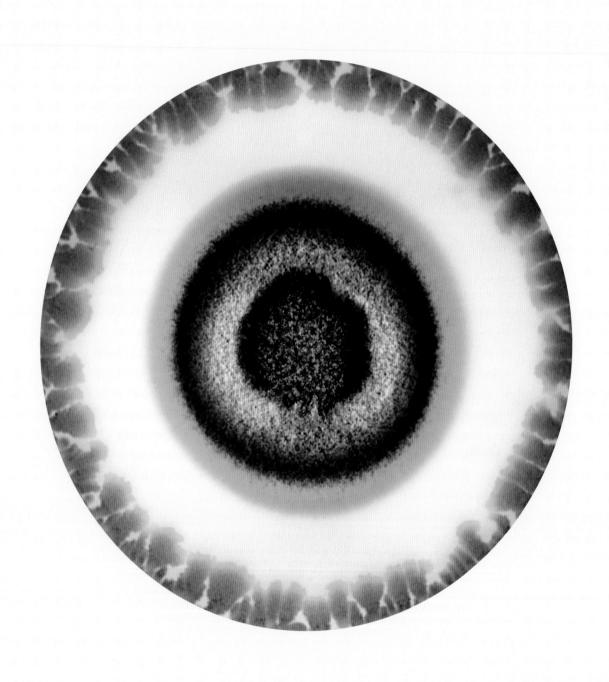

Penicillium notatum: A fungus commonly found in nature and selected for its capacity to produce the antibiotic penicillin. Penicillin kills adjacent bacteria, producing the empty halo you can see around the colony in the facing photograph.

It would take more than 5,600,000,000 *Penicillium notatum* to fill just one sombrero.

Poliovirus

Poliomyelitis (Po·lee·oh·my·ill·eye·tis) **POLIO**

Long ago, summertime, with no more school,
There was lots of time in the swimming pool

For Rog, Tom, and Ken, brothers three,
To play and frolic, trouble free.

But not so glad were Dad and Mum.
The fear of polio made them glum.

They'd worry and fret at each sneeze or wheeze,
Since with summer came this dread disease.

One year earlier, nineteen fifty-four,
Polio struck at Steve, who lived next door.

Just a cold one day, but the very next
Came a headache no aspirin could arrest.

His muscles stiffened, both legs and arms,
And troubled breathing brought real alarm.

Thank goodness a lung machine kept him alive,
And with nursing care he did survive.

But Stevie's legs were paralyzed,
Strengthened little by exercise.

"What," asked Rog, "is **poliovirus**?
How does it get so deep inside us?"

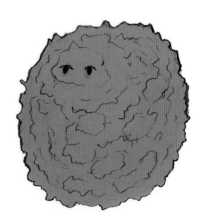

Polio swallowed first makes its way
Into the blood, then into nerves to stay.

A virus is so very small
It cannot thrive alone at all.

Inside a nerve cell it can multiply
Until thousands make the nerve cell die.

Muscles wither when nerves are dead
Anywhere from feet to head.

"But what about poor Steve?" asked Ken.
"Can **poliovirus** get him again?"

No, because he's made antibodies

That trap the virus and stop the disease.

"We want antibodies," young Tom pleaded,

"To fight off the virus when they're needed."

Don't fret, kids, polio can now be beat

With a virus vaccine that is pretty neat.

"What's a vaccine?

Does it make you sick?

Beating polio must be quite a trick."

In a vaccine, germs are made so weak

Body cells can remove them without a squeak.

Thus cells learn to make antibodies

That can dispose of **poliovirus** as quick as you please.

Hail polio vaccine, gift of science!
With it, toward polio, we show defiance!

Were this vaccine given to everyone
The polio battle would be forever won.

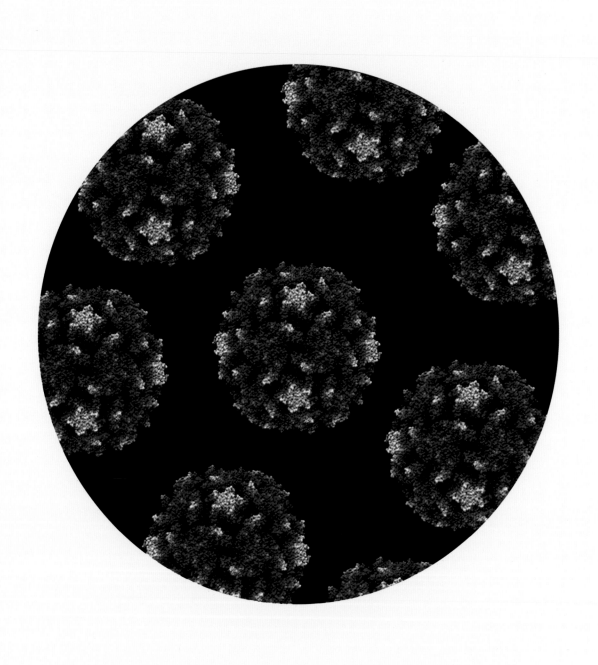

Polio: A short name for the virus that causes poliomyelitis. This epidemic disease, preventable by a safe vaccine, may cause muscular paralysis and atrophy. The facing photograph shows particles of poliovirus models based on crystallographic studies.

Roughly 6,700,000,000 particles of poliovirus could fit on the wing of a fly.

Human Immunodeficiency Virus AIDS

At Portola School, in the second grade,

Kids whispered about a disease called AIDS.

This dread disease affected Bill—

Yet he didn't seem all that ill.

Concerned Zoe asked her mom and dad

Just what it was her schoolmate had.

"Why call it AIDS? It does no good

As Band-Aids or reading aids would or should."

AIDS is simply an abbreviation

For an immune system aberration.

Acute *ImmunoDeficiency*

Syndrome together spell AIDS, you see.

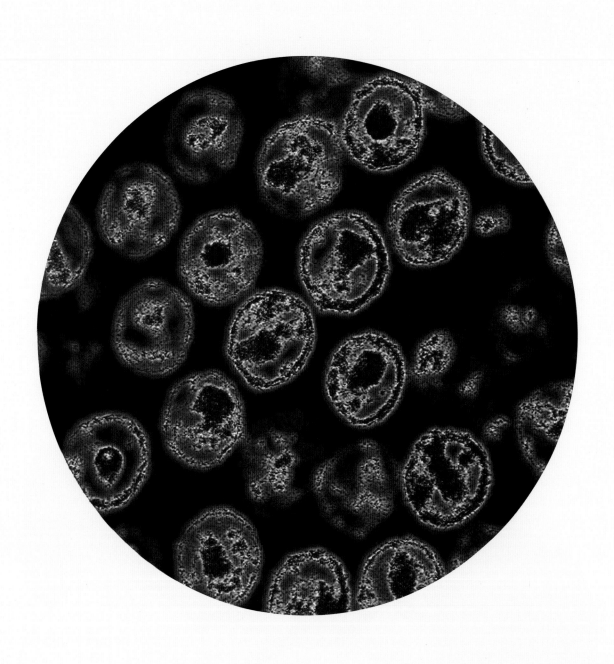

Human Immunodeficiency Virus AIDS

At Portola School, in the second grade,
Kids whispered about a disease called AIDS.

This dread disease affected Bill—
Yet he didn't seem all that ill.

Concerned Zoe asked her mom and dad
Just what it was her schoolmate had.

"Why call it AIDS? It does no good
As Band-Aids or reading aids would or should."

AIDS is simply an abbreviation
For an immune system aberration.

Acute *ImmunoDeficiency*
Syndrome together spell AIDS, you see.

The AIDS virus is called **HIV**

For *H*uman *V*irus of *I*mmunodeficiency.

"Why do kids from Bill stay away

In class and in the yard at play?"

Folks think that AIDS can infect you

Like mumps or measles, pox or flu:

By sneeze, cough, or touch, as viruses are shed.

But that's not the way that AIDS is spread.

Though kids bite and shove and scratch every day,

No one ever catches AIDS that way.

"How did Bill get **AIDS**? And from where?

How kids avoid him seems so unfair."

Bill has hemophilia, as you know.

He bleeds because his clotting's slow.

At times he needs a blood transfusion

To replace the blood he's too often losin'.

Years ago he was sadly fated

To receive blood that was contaminated

With AIDS virus from someone infected.

Tests back then could not detect it.

"AIDS sounds so bad.

Is there a vaccine or cure?"

Not long ago Bill's chances were poor.

But with research advances, drugs now have a role

To keep **HIV** under firm control.

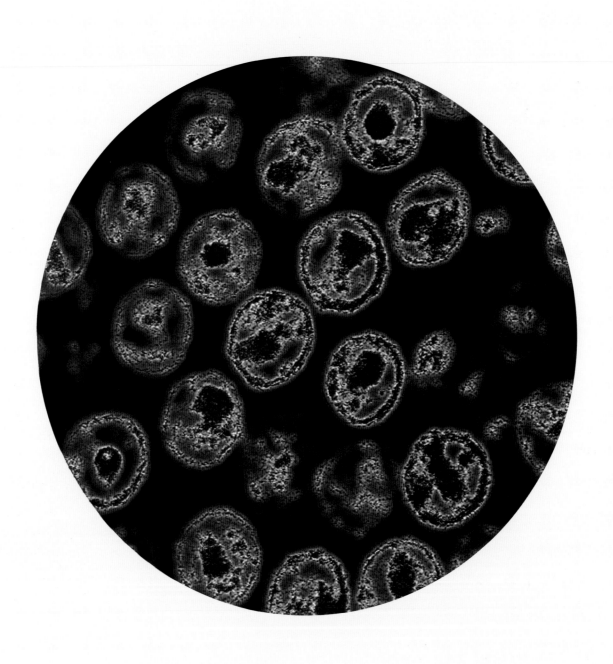

HIV: The abbreviation for human immunodeficiency virus, the agent that causes acute immunodeficiency syndrome, or AIDS, in which the victim lacks the ability to combat invasive germs. The facing electron micrograph shows a group of AIDS virus particles.

A measly grain of sand is 8,000,000,000 times bigger than a single particle of HIV.

The Intestinal Menagerie

In the intestines of all of us,
Infants and old,
Dwells a bacterial menagerie
With creatures trillion-fold.

Soon after our birth, the bugs multiply
Into hundreds of kinds, all diversified.

Maya then asked, "Are bacteria in me
Different from ones in my brother Gili?"

Yes, each person's menagerie is unique—
Just as our fingerprints are, so to speak.

These bugs make vitamins and digest food,
And their great number
Doesn't let bad germs intrude.

But how the bugs digest food

May make some people obese

And also might make

Some diseases increase:

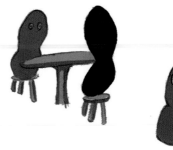

Heart disease, high blood pressure, and diabetes,

Cancer, gallstones, and even lung disease.

These days the environment is much in the news,

How we should protect it and prevent its abuse.

Global warming is something that we now must fear.

Melting ice caps and flooding seem all too near.

What gets less attention

Yet matters a great deal

Is the environment inside us

That we may not feel.

Hence, tales of germs and

Our bacterial menagerie within

Need telling to teach us

Why we have been

Ill or well—

Or even fat or thin.

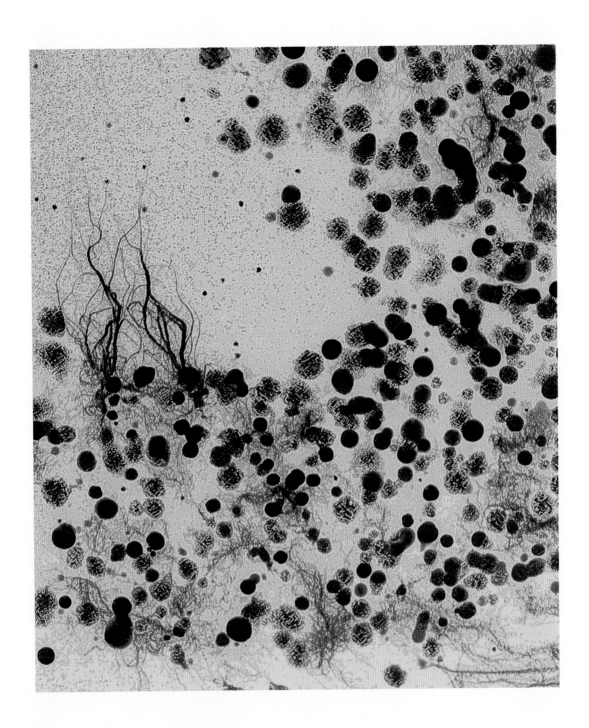

A Bacterial Menagerie

This photograph shows many species
of bacteria growing on a surface.

Pyrococcus furiosus: Named Rushing Fireball (*Pyrococcus furiosus*) because it swims very rapidly at temperatures above 80° C (176° F), this germ has as many as 70 flagella, which it uses for swimming and for sticking to surfaces to make a biofilm.

About 354,880,000,000,000 of these Rushing Fireballs could fill a jelly jar.

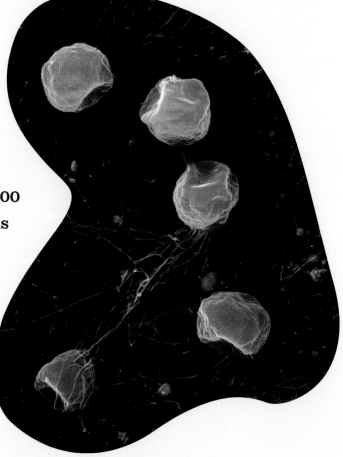

Glossary

Agar A semisolid medium used to grow a bacterial culture.

Antibiotics Chemicals released from many kinds of microbes (generally soil dwellers) that interfere with the growth of other microbes; the huge variety includes penicillin, cephalosporin, and streptomycin.

Antibody One of millions of slightly different kinds of protein molecules produced by the body, each kind multiplied in response to a specific foreign substance or cell, such as a toxin, pollen, or microbe; by binding that foreign entity, the antibody helps neutralize its action and remove it.

Antitoxin An antibody produced in the body in response to a specific toxin; by binding that toxin, it helps neutralize its action.

ATP The abbreviation for adenosine triphosphate, the currency of energy metabolism—when cells burn food, such as glucose, some of the energy is captured as ATP and used for muscle contraction, vision, and other body functions.

Bacteria (*singular: bacterium*) The smallest cell, each generally 1/1,000 the size of a plant or animal cell; also called a germ.

Biofilm A community of bacteria that adhere to one another in a sheet-like structure; more than 90 percent of bacteria in nature exist in biofilms rather than as individuals.

Carrier A person who carries disease germs to which he or she is immune but from whom the germs can be spread.

Cell The fundamental unit of life that can live and reproduce on its own but often is organized into a plant or animal tissue or a microbial colonial mass.

Cephalosporin A commonly used antibiotic drug.

Clostridium tetani (klos·tri·dee´·um tet´·an·eye) A rodlike bacterium that can become a spore to attain a highly resistant dormant state; introduced through a wound, the spore can germinate to become a bacterium again, releasing toxins that cause severe muscle contraction in the arms, legs, diaphragm, and jaw.

Coat The wall around a bacterial cell.

Colony A mass of bacteria in the millions growing on a semisolid surface, sometimes starting from a single germ.

Crystallography The study of the structure of molecules in their ordered crystalline form.

Diabetes A disease commonly associated with excessive levels of sugar in the blood and urine due to the lack of insulin, a hormone produced in the body.

Dormant A resting state in the life of a bacterium.

Electron micrograph An image magnified a thousandfold compared to that seen in an ordinary microscope.

This photograph shows a multitude
of microbes growing as a biofilm on a
rock's surface in a pond at Yellowstone.

Enzyme A class of chemicals produced by cells that accelerates all the chemical operations in the body, such as digestion of food.

Esophagus The tube that leads from the throat to the stomach, which in humans is about nine inches long.

Flagella (*singular: flagellum*) The long, thin appendages of bacteria that are responsible for their motion.

Helicobacter pylori (hee´·li·ko·bak·ter pie·law´·ree) A helix-shaped, common bacterium that screws into the lining at the pit of the stomach (the pylorus); it releases an enzyme that enables it to neutralize corrosive stomach acid. In a few individuals it creates an open sore, a peptic ulcer, that may progress to cancer.

Hemophilia A familial disease marked by a tendency to bleed excessively because of the blood's inability to clot.

HIV The abbreviation for *human immunodeficiency virus*, the agent that causes AIDS, in which the victim lacks the ability to combat invasive germs.

Lobe Part of an animal organ, as of a lung or brain.

Mat A solid growth of bacteria on a semisolid surface composed of billions of microbes.

Microbicide An agent that kills microbes (germs).

Microscope An optical instrument composed of convex lenses through which an illuminated cell is magnified up to 1,000-fold; to observe viruses and fine cellular details, an object is magnified up to 100,000-fold by electron bombardment in an electron microscope.

Mold A fungus in a very large and varied group of microbes that includes bacteria, such as *Myxobacteria*.

Molecule The smallest chemical particle made up of one or more atoms (for example, water as H_2O).

Mucous Covered with mucus or slime.

Mutant / Mutation An alteration in a gene that may give the bacterium a different property or appearance.

Mycoplasma mobile (my·ko·plaz´·ma mo·bee´·lay) A small germ that causes disease. (See page 69.)

Myxobacterium xanthus (miks·o·bac·teer´·ee·um zan´·thus) One of many species of complex bacteria that produce slime on which they glide to find and feed on other bacteria and decaying matter.

Penicillium notatum (pen·i·sill´·ee·um no·tah´·tum) A fungus commonly found in nature and selected for its capacity to produce the antibiotic penicillin.

Peptic ulcer A break in the lining of the stomach or small intestine that can be a persistent, bleeding sore.

Poliovirus A short name for the virus that causes poliomyelitis (po´·lee·oh·my·ill·eye´·tis); this epidemic disease, preventable by a safe vaccine, may cause muscular paralysis and atrophy.

Poly P A *poly*mer of *p*hosphates, each made of a phosphorus atom surrounded by and linked to others by oxygen atoms. Poly P chains, hundreds long, were likely involved in the origin of life on earth and carry out important functions in every cell in nature—bacteria, fungi, plants, and humans.

Protein A major and essential component of all cells that is made up of amino acids, which in turn are made up of the elements carbon, hydrogen, nitrogen, and oxygen.

Pustule A small skin elevation containing inflammatory cells and perhaps infectious agents.

Pyrococcus furiosus (pie·ro·kok´·us fu·ree·o´·sus) A bacterium that dwells in hot springs. (See page 64.)

Resistant In the case of bacteria, a state in which they resist an antibiotic or the immune system of the host.

Rod A bacterium shaped like a little twig.

Saccharomyces cerevisiae (sa·kar´·o·my·sees ser·a·vee´·see·yay) A yeast fungus (*see* Yeast) found in fruit or soil or on human skin; it is widely used in baking bread and brewing.

Salmonella typhi (sal·ma·nel´·a tie´·fee) A rodlike bacterium related to the many bacteria that normally reside in the intestines; the toxins it produces invade the bloodstream to cause typhoid fever.

Spore A dormant form of microbe surrounded by many coats that make it resistant to heat, drying, and chemicals but capable even after many years of being awakened (germinated) to a living cell.

Stalk A stem-like organization of a bacterial population.

Staphylococcus aureus (staf·lo·kok´·us or´·ee·us) A spherical bacterium commonly found on skin and mucous membranes; its grapelike clusters, which appear golden as a colonial mass, can produce toxins that invade the body and cause serious infections.

Streptococcus pneumoniae (strep·to·kok´·us noo·mo´·nee·ay) A small, round bacterium found in pairs or short chains; it causes serious human diseases such as lobar pneumonia.

Toxin A poisonous substance produced by a germ that in animals can give rise to specific antitoxins as an immunity defense.

Vaccine A preparation of killed or weakened germs administered to animals and humans in order to stimulate production of defenses, such as antibodies, that protect against that particular germ.

Vesicle A compartment within a cell.

Virus A subcellular form as small as 1/1,000 the size of a bacterium; it must parasitize or live in a cell (bacterial, plant, or animal) for nourishment and reproduction.

Yeast A microbe, larger and more complex than bacteria, that is related to other fungi and molds; it resembles plant and animal cells and reproduces by fission or budding. Yeast is used in baking and brewing (see *Saccharomyces cerevisiae*).

Mycoplasma mobile: On solid surfaces such as glass, *Mycoplasma mobile* can glide by using its special "leg" and "foot" (yellow and red in the photograph below).

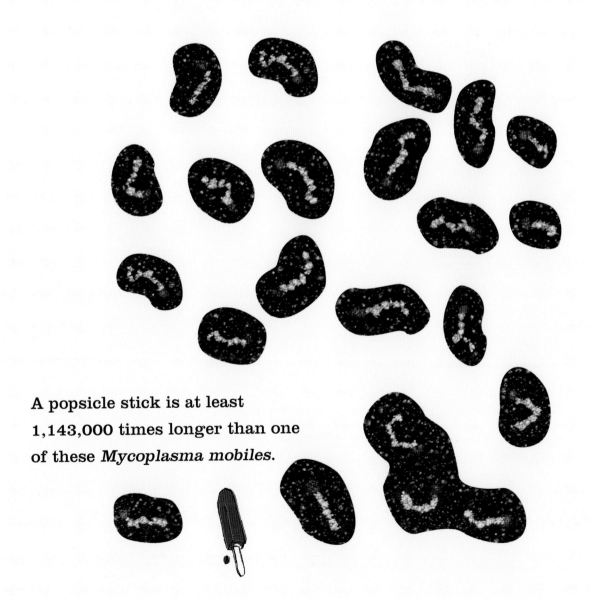

A popsicle stick is at least 1,143,000 times longer than one of these *Mycoplasma mobiles.*

A Note on the Photography

Images from the microbial world must be obtained in many ways, since the sizes of the objects under observation can vary enormously. Microbial colonies on a plate or on rocks in a pond, measuring from a few millimeters to many meters, can easily be captured with a regular camera using a normal lens or a close-up lens. But to create images of much smaller things, such as individual cells, a camera attached to a light or electron microscope is necessary. Light microscopy works with objects as small as one millionth of a meter. Still smaller things can be captured by an electron microscope. The resulting images, which are usually gray, can be artificially colored. This process gives particularly interesting features more contrast and makes them more visible. The resulting photographs can be quite beautiful.

All of the photographs in this book, from the sources listed below, were edited and color enhanced by Dr. Roberto Kolter using Adobe Photoshop. Dr. Kolter grew up in Guatemala and moved to the United States to attend school at Carnegie Mellon University; the University of California, San Diego; and Stanford University. Since 1983 he has been professor of microbiology and molecular genetics at Harvard Medical School. Always interested in obtaining and presenting images from the microbial world, he initiated the tradition of illustrating the cover of the *Journal of Bacteriology*. He has been that journal's cover editor since 1999, and many of the photographs in this book previously appeared as cover images for that journal.

The following photographs were created by Dr. Kolter: pages viii, 15, 32, 46, 52, 58, 63, 66, and back cover.

Other photographs in this book were made by R. P. Ross, page 8; E. Peter Greenberg and Jeremy Yarwood, page 9; Eduardo Groisman, page 14; Adam Driks, page 20; Michael Thumm, page 26; David Engelberg, page 27; Ernesto Garcia, page 33; Dale Kaiser, page 36; Yves Brun and David White, page 37; Reinhard Wirth, page 64; and Makoto Miyata, page 69. The image on page 41 is courtesy of www.hpylori.com.au.

On the way to Stockholm, 1959: Rog, Ken, Sylvy, Dr. Kornberg, and Tom.

About the Author

During his early career as a research enzymologist, Arthur Kornberg began telling his sons, Rog, Tom, and Ken, his charming "germ stories" about the heroes and baddies of the germ world and the exploits of the scientists who studied them.

Kornberg won the 1959 Nobel Prize in Physiology or Medicine for discovering the role enzymes play in the replication of deoxyribonucleic acid (DNA). In that same year he founded the Department of Biochemistry at Stanford University School of Medicine, where he has continued to teach and to research topics such as DNA synthesis, DNA replication, and inorganic polyphosphate (poly P).

Just as Rog, Tom, and Ken figured in the earlier "germ stories," their children, Gili, Guy, Jessica, Maya, Ross, Sophie, Zac, and Zoe, have become the catalysts and eager audience for a new set of tales; their questions and adventures, along with their grandfather's lifetime of scientific research, bring to life the "little beasties" and the microscopic world they inhabit.

Clockwise, from left: Jessica, Sophie, Ross, Guy, Zac, Maya, Gili, and Zoe.

A Note on the Illustrations

To create the whimsical and humorous drawings of germs that appear throughout this book, Adam Alaniz researched scholarly texts and studied many photographs and electron micrographs of bacteria. He used watercolor, pen, and ink on Arches paper for the spot illustrations and a combination of watercolor, pen, ink, paper, and computer enhancement for the larger full-page illustrations. Mr. Alaniz, who has been drawing since the third grade, graduated from the Art Center College of Design in Pasadena, California.

Produced by Wilsted & Taylor Publishing Services

Project management: Christine Taylor

Art management: Jennifer Uhlich

Production assistance: Drew Patty and Mary Lamprech

Copyediting: Melody Lacina

Proofreading: Nancy Evans

Design and composition: Tag Savage

Printer's devils: Lillian Marie Wilsted, Juna Hume Clark,
Gracie Quinn, and Annie Quinn

Printing and binding: Regal Printing Ltd., Hong Kong,
through Stacy and Michael Quinn of
QuinnEssentials Books and Printing, Inc.